Land Your Dreams
Volume Two
By: I. Wright

I. Wright

Copyright © 2024 by I. Wright

All rights reserved. No part of this book may be reproduced or transmitted in any form or by any means, electronic or mechanical, including photocopying, recording, or by any information storage and retrieval system, without permission in writing from the publisher.

For permission requests, please email:
thewyldernessprogram@gmail.com

ISBN: 979-8-9880644-1-1

BACK HOUSE
PUBLISHING

I. Wright

Land Your Dreams Volume Two

Table of Contents

Dedication..*v*

Introduction *1*

Manifesting Your Dreams: A Journey of Purpose and Mindset ..*2*

Challenges of Packing Up a House After 20 Years *5*

Preparing to Build on Rural Land...*7*

Working with Your Local USDA Office *9*

USDA Programs and Services *11*

General Tips for Rural Living *12*

Main Components and Their Associated Costs: *15*

Living Off-Grid and Starting a Garden *21*

Caring For A Garden *27*

Root Cellar *30*

Bunkers *32*

Water Access *34*

About The Author *40*

Notes..*42*

I. Wright

Land Your Dreams Volume Two

Dedication

I dedicate this book to my son Abdur-Rahim Lamour Robinson (ALR). Semi-living off the grid is something you always spoke about. Your carefree lifestyle will live with us forever. Coming from a big city, living in the south you've learned to hunt and have chickens of your very own. When we all laughed at you having your own chickens, I would've never seen myself doing the same. You inspired me and I didn't even see it! I thank my ancestors for allowing me to be your mother for 24 years. Your legacy will live on.

I. Wright

Introduction

Embarking on the journey of building on rural land and transitioning to an off-grid lifestyle can be both exhilarating and challenging. This guide is designed to equip you with essential knowledge and practical steps to ensure a successful and smooth transition. Every detail is crucial from the initial preparations of obtaining permits, conducting soil tests, and setting up utilities, to the emotional and physical demands of packing up a house after 20 years.

Furthermore, understanding the principles of manifesting your dreams and working closely with your local USDA office can provide invaluable support.

This guide also addresses the realities of living off the grid, highlighting both the advantages and disadvantages to help you make informed decisions. By embracing a mindset of purpose and aligning your thoughts, beliefs, and actions, you can manifest your dreams and create a fulfilling rural lifestyle.

Manifesting Your Dreams: A Journey of Purpose and Mindset

I had a life-changing meeting with my children about buying land, which sparked a year-long journey towards making our dream a reality. From that moment, I was determined and focused, acting as if we were already living on the land we envisioned. Throughout the year, I never wavered in my belief that our dream would come true, constantly affirming our aspirations. This book is a testament to the power of purpose and mindset, designed to guide you on your journey of understanding and embracing your purpose. It teaches you how to cultivate a mindset that aligns with your desired reality, enabling you to manifest your dreams into tangible results.

Steps to Manifest Your Dreams

1. **Define Your Desires**: Reflect on what brings you joy and fulfillment, and write down your dreams in a clear, specific manner using positive language.

2. **Align Thoughts, Beliefs, and Actions**: Cultivate a mindset of abundance and positivity. Visualize yourself living your dream and feel the associated emotions.

3. **Take Inspired Action**: Break down your goals into manageable steps and take consistent action towards them.

4. **Surround Yourself with Support**: Engage with positive and supportive people who believe in your dreams.

5. **Patience and Trust**: Trust the process and stay committed, even in the face of challenges. Maintain a positive attitude and use challenges as opportunities for growth.

6. **Visualization and Affirmations**: Daily visualization and affirmations help create a strong mental image and align your thoughts with your desired outcome.

7. **Gratitude**: Cultivate gratitude for what you already have and the progress you've made.

8. **Positive Self-Talk**: Replace negative thoughts with positive and empowering statements.

9. **Self-Care**: Take care of your physical, mental, and emotional well-being to create a supportive environment for manifesting your dreams.

10. **Journaling**: Keep a journal to document your journey, thoughts, and progress. This helps in maintaining focus and reflecting on your growth.

11. **Mindfulness and Meditation**: Incorporate mindfulness and meditation practices to stay grounded and focused on your goals.

12. **Learning and Growth**: Continuously seek knowledge and skills that align with your dreams. This could include taking courses, reading books, or attending workshops.

By aligning your thoughts, beliefs, and actions, and staying committed to your vision, you can manifest your dreams into reality.

Remember, if you can think it, you can have it. Once your mindset shifts from your current situation to the reality you envision, it becomes yours to walk in. So, embrace the teachings in this book, apply them to your life, and confidently step into your purpose. Your dreams are within reach, waiting for you to claim them.

Challenges of Packing Up a House After 20 Years

Packing up a house after living there for over 20 years comes with its fair share of challenges:

1. **Decluttering**: Deciding what to keep, donate, sell, or discard among years of accumulated belongings can be time-consuming and emotionally challenging.

2. **Time and Organization**: Efficiently packing a large number of belongings requires careful planning and time management, especially when balancing other responsibilities.

3. **Physical Demands**: Packing involves heavy lifting and carrying, which can be physically demanding. Proper lifting techniques and seeking assistance can help avoid injury.

4. **Emotional Attachment**: Letting go of items with sentimental value can be emotionally challenging. Allow yourself time to process these emotions.

5. **Logistics and Coordination**: Coordinating with movers, arranging transportation, and managing utility transfers require careful planning and attention to detail.

6. **Space Constraints**: Finding adequate storage space for packed boxes may necessitate creative solutions, such as renting storage units.

7. **Repair and Maintenance**: Addressing repairs and maintenance tasks before moving out can add complexity and time to the packing process.

8. **Professional Help**: Consider hiring professional organizers or moving companies to assist with the packing and moving process. They can provide expertise and efficiency.

9. **Inventory**: Create an inventory of your belongings. This helps in organizing, packing, and later, in settling into the new place.

10. **Labeling**: Clearly label all boxes with their contents and the room they belong to. This simplifies the unpacking process.

Approaching the packing process systematically and seeking assistance when needed can help overcome these challenges. Prioritize self-care and celebrate your progress along the way.

Preparing to Build on Rural Land

There are several crucial steps that will ensure a smooth and successful process:

1. **Obtain Permits**: Start by obtaining permits from the zoning department in your county to ensure your construction plans comply with local regulations.

2. **Soil Testing**: Conduct a soil test before drilling a well or installing a septic system. This determines the suitability of the soil for construction, agriculture, and waste management.

3. **Electric Power**: Determine the location of the nearest electric power line if you don't plan to use solar power. Contact the electric company to inquire about the cost of connecting your land to the power grid.

4. **Water Supply**: Plan for a well, which can cost around $12,800 depending on factors like soil type (rock or sand). Research and secure water rights, especially in areas where water resources are scarce or regulated.

5. **Septic System**: Every home requires its own septic system, with costs ranging from $7,000 to $11,000 depending on the home's size.

6. **Grading**: Before building, or if considering a manufactured home, you'll need to grade the area. Grading costs typically start at $5,000 and can vary based on the home's size.

7. **Manufactured Homes**: If opting for a manufactured home, clarify whether the installation is included in the home's price or needs to be arranged separately.

8. **Internet and Communication Services**: Check the availability of internet and communication services in the area. Consider satellite options if traditional services are not available.

9. **Insurance**: Look into obtaining appropriate insurance for your rural property, including coverage for natural disasters specific to the area, such as floods, wildfires, or earthquakes.

10. **Financing**: Some mortgage companies may finance the well and septic system but may require your land as collateral. It's generally recommended to keep your land

separate from financial obligations to protect your ownership rights.

11. **Safety and Privacy**: Consider fencing and a security system to enhance the safety and privacy of your property. Building on rural land can be costly, but with careful planning and budgeting, it can be an affordable and rewarding endeavor. –

Working with Your Local USDA Office

To ensure you have a farm number and access to USDA programs, contact your local USDA office. They provide essential resources and guidance for farmers, including:

1. **Programs and Services**: Financial assistance, crop insurance, conservation programs, and market access programs.

2. **Farm Planning and Guidance**: Expert advice on crop selection, soil conservation, pest management, and livestock care.

3. **Regulatory Compliance**: Assistance with understanding and complying with agricultural regulations.

4. **Market Access and Trade Opportunities**: Information on export opportunities, certifications, and navigating trade regulations.

5. **Networking and Community Building**: Connecting with other farmers, agricultural organizations, and industry stakeholders.

USDA Programs and Services

1. **Farm Loans**: Loans for purchasing land, equipment, livestock, and operating expenses with favorable terms.

2. **Crop Insurance**: Financial protection against crop losses due to natural disasters, pests, and other risks.

3. **Conservation Programs**: Financial assistance and technical expertise for sustainable farming practices.

4. **Market Access Programs**: Assistance with market research, trade shows, export promotion, and certifications.

5. **Research and Education**: Funding for agricultural research, education, and extension services.

6. **Food Safety and Inspection Services**: Ensuring the safety and quality of the food supply through inspections and guidance.

7. **Rural Development Programs**: Funding for infrastructure projects, rural housing assistance, and business development.

8. **Eligibility Criteria**: Familiarize yourself with the eligibility criteria for various USDA programs to ensure you qualify and prepare necessary documentation.

9. **Application Process**: Understand the application process for USDA programs. This includes knowing deadlines, required documents, and potential waiting periods.

10. **Community Involvement**: Get involved with local agricultural communities or associations. This can provide networking opportunities, support, and additional resources.

Working closely with your local USDA office can be a key resource for your farm's success and sustainability. Contact your local USDA office or visit the USDA website for more information on available programs.

General Tips for Rural Living

1. **Emergency Preparedness**: Prepare for emergencies by having a first aid kit, emergency food and water supplies, and a plan for evacuation if needed.

2. **Sustainability Practices**: Consider implementing sustainable practices such as rainwater harvesting, composting, and renewable energy sources to reduce environmental impact and improve self-sufficiency.

3. **Local Resources**: Identify local resources and services, such as nearby hospitals, fire departments, and veterinary services, to ensure you have access to essential services.

Living off the grid has its advantages, but it also comes with a few disadvantages.

Some of the disadvantages include:

1. **Initial setup costs**: The initial investment required to set up an off-grid system can be significant. Installing solar panels, wind turbines, batteries, and other renewable energy systems can be costly.

2. **Limited access to utilities**: Living off the grid means being self-sufficient, which also means limited access to utilities such as water, electricity, and sewage. You may need to rely on alternative sources like rainwater harvesting, solar power, and composting toilets.

3. **Maintenance and troubleshooting**: Off-grid systems require regular maintenance and troubleshooting. This includes maintaining solar panels, batteries, and other equipment. If something goes wrong, you may need to troubleshoot and fix it yourself or hire a professional, which can be time-consuming and costly.

4. **Limited amenities and conveniences**: Living off the grid often means sacrificing certain amenities and conveniences that are readily available in urban areas. For example, you may have limited internet access, limited entertainment options, and limited access to certain goods and services.

5. **Reliance on weather conditions**: Off-grid systems rely heavily on weather conditions. If there is not enough sunlight or wind, the production of electricity may be limited. This can impact your ability to power appliances and other devices.

6. **Social isolation**: Living off the grid can sometimes lead to social isolation, especially if you are living in a remote area. Limited interaction with neighbors and reduced access to community resources can make it challenging to maintain a social life.

Considering these disadvantages and weighing them against the benefits is important before deciding to live off the grid. The initial setup costs for an off-grid system can vary depending on various factors such as the system's size, the household's energy requirements, and the location.

Main Components and Their Associated Costs:

1. **Solar panels**: The cost of solar panels can range from a few hundred dollars to several thousand dollars, depending on the size and quality of the panels. The average cost for a residential solar panel system can be around $10,000 to $20,000 or more.

2. **Wind turbines**: If you are considering using wind power, the cost of a wind turbine can range from a few thousand dollars to tens of thousands of dollars, depending on the size and capacity of the turbine.

3. **Batteries**: Off-grid systems require batteries to store excess energy for use during times when renewable energy sources are not producing enough power. The cost of batteries can vary depending on the capacity and type, but

they can range from a few hundred dollars to several thousand dollars.

4. **Charge controllers and inverters**: Charge controllers are used to regulate the charging of batteries. In contrast, inverters are needed to convert the DC power from the batteries into AC power for household use. The cost of charge controllers and inverters can range from a few hundred dollars to a few thousand dollars.

5. **Installation and labor costs**: It's important to consider the cost of installation and labor when setting up an off-grid system. Depending on the complexity of the installation and the need for professional help, this can add a significant cost to the overall setup.

It's worth noting that while the initial setup costs can be high, off-grid systems can provide long-term savings on utility bills and reduce reliance on traditional energy sources.

Additionally, there may be government incentives or tax credits available to help offset some of the initial costs. Yes, there are various government incentives and tax credits available in many countries to help offset the initial costs of setting up an off-grid system. These incentives and credits are designed to promote the use of renewable energy and encourage individuals and businesses to invest in clean energy solutions.

Here are some examples:

1. **Federal Investment Tax Credit (ITC):** In the United States, the federal government offers a tax credit for residential and commercial solar installations. The ITC allows homeowners and businesses to deduct a percentage of the cost of their solar system from their federal taxes. As of 2021, the ITC provides a 26% tax credit for qualifying solar installations.

2. **State and Local Incentives**: Many states and local governments offer additional incentives and rebates for going solar or installing other renewable energy systems. These incentives can vary widely and may include cash rebates, grants, property tax exemptions, or sales tax exemptions. It's advisable to check with your state or local authorities to see what incentives are available in your area.

3. **Net Metering**: Net metering is a policy that allows homeowners and businesses with renewable energy systems to receive credit for any excess electricity they generate and feed back into the grid. This credit can offset the cost of electricity consumed when the renewable

energy system is not producing enough power. Net metering policies vary by location, so it's important to consult with your utility company or local authorities to understand the specific regulations and benefits in your area.

4. **Renewable Energy Certificates (RECs):** Some regions have programs that offer Renewable Energy Certificates for electricity generated from renewable sources.

These certificates represent the environmental benefits of clean energy generation and can be bought or sold. By purchasing RECs, individuals and businesses can support renewable energy projects and offset their own energy consumption.

It's important to note that the availability and specific details of government incentives and tax credits can vary depending on your location. It's recommended to consult with a local renewable energy expert or reach out to your government's energy department to get accurate and up-to-date information on the incentives available in your area.

The Federal Investment Tax Credit (ITC) in the United States is a tax credit offered by the federal government to incentivize the adoption of renewable energy systems, including solar energy. The ITC allows residential and commercial property owners to deduct a percentage of the cost of their solar system from their federal taxes.

As of 2021, the ITC provides a 26% tax credit for qualifying solar installations. This means that if you install a solar energy system on your property, you can deduct 26% of the total system costs from your federal tax liability.

For example, if your solar system costs $20,000, you could potentially receive a tax credit of $5,200 (26% of $20,000). It's important to note that the ITC is a credit, not a deduction. This means that the tax credit directly reduces the amount of taxes you owe, rather than reducing your taxable income.

If the tax credit exceeds your tax liability for the year, you may be able to carry the remaining credit forward to future years. The ITC has played a significant role in promoting the growth of solar energy in the United States by making solar installations more affordable for homeowners and businesses.

However, it's essential to consult with a tax professional or visit the official website of the Internal Revenue Service

(IRS) to understand the specific eligibility criteria and requirements for claiming the ITC, as they may be subject to change or vary based on individual circumstances.

Living Off-Grid and Starting a Garden

Here are some first steps to consider:

1. **Assess Your Land**: Evaluate the available space on your property for gardening. Consider factors such as sunlight exposure, soil quality, and drainage. Choose an area that receives ample sunlight for most of the day and has fertile soil.

2. **Plan Your Garden**: Determine the size and layout of your garden. Consider the types of plants you want to grow, their space requirements, and any companion planting strategies you may want to employ. Sketch out a garden design to help visualize the layout.

3. **Prepare the Soil**: Clear the designated area of any weeds, rocks, or debris. Loosen the soil using a gardening fork or tiller to improve its texture and allow for better root penetration. Add organic matter like compost or well-rotted manure to enrich the soil.

4. **Choose Your Plants**: Select plants that are suitable for your climate, soil type, and available space. Consider both vegetables and herbs that are easy to grow and provide a

good yield. Start with a mix of seeds and seedlings to give your garden variety.

5. **Planting**: Follow the planting instructions for each plant, considering factors such as spacing, depth, and watering requirements. Plant seeds or transplant seedlings into the prepared soil, ensuring they have enough room to grow.

6. **Watering**: Water your plants regularly, especially during dry periods. Use a watering can, hose, or drip irrigation system to provide consistent moisture. Be mindful not to overwater, as it can lead to root rot or other issues.

7. **Weed Control**: Regularly remove weeds that compete with your plants for nutrients and resources. Use mulch or organic weed control methods to suppress weed growth and conserve moisture in the soil.

8. **Pest Management**: Monitor your garden for pests and take appropriate measures to control them. Consider organic pest control methods such as companion planting, natural predators, or homemade remedies.

9. **Maintenance**: Regularly maintain your garden by pruning, staking, and providing support for plants as

needed. Remove dead or diseased plant material to prevent the spread of diseases.

10. **Harvest and Enjoy**: Once your plants have matured, harvest your crops and enjoy the fruits of your labor. Preserve any excess produce through canning, freezing, or drying for future use.

Remember, gardening off-grid may require additional considerations, such as water conservation, alternative energy sources for irrigation, and sustainable practices. Adapt your gardening methods to suit your off-grid lifestyle and continue learning and experimenting as you gain experience.

To start your garden, you will need the following:

1. **Seeds or plants**: Decide what type of plants you want to grow and purchase the seeds or young plants accordingly. Consider factors such as your climate, available space, and personal preferences.

2. **Soil**: Choose a high-quality soil that is suitable for the plants you want to grow. Different plants have different soil requirements, so make sure to research and choose the right type of soil for your garden.

3. **Containers or garden beds**: Decide whether you want to grow your plants in containers or in a traditional garden bed. If you choose containers, make sure they have drainage holes and are the right size for your plants. If you opt for a garden bed, prepare the area by removing any weeds or grass and loosening the soil.

4. **Watering tools**: To keep your plants hydrated, you will need watering tools such as a watering can or hose. Consider the size of your garden and choose the appropriate watering tools.

5. **Gardening tools**: Some essential gardening tools include a garden trowel, hand rake, pruning shears, and a garden fork. These tools will help you with planting, weeding, pruning, and other maintenance tasks.

6. **Fertilizer**: Depending on the type of plants you are growing, you may need to use fertilizers to provide essential nutrients to your plants. Choose organic or synthetic fertilizers based on your preferences and the needs of your plants.

7. **Sunlight**: Most plants require adequate sunlight to grow and thrive. Identify the areas in your garden that receive sufficient sunlight and plan your plant placements accordingly.

8. **Protection**: Consider using protective measures such as fencing or netting to keep pests and animals away from your garden. You may also need to protect your plants from extreme weather conditions, such as using row covers during frost or providing shade during hot summer days.

Remember to research the specific needs of the plants you want to grow and adapt your gardening approach accordingly. Happy gardening!

Researching the specific needs of the plants you want to grow is important for several reasons:

1. **Proper care**: Different plants have different requirements in terms of sunlight, water, temperature, soil type, and nutrients. By researching the specific needs of your chosen plants, you can ensure that you provide them with the optimal conditions for growth. This will help your plants thrive and produce better results.

2. **Preventing diseases and pests**: Some plants are more susceptible to certain diseases and pests than others. By understanding the specific needs and vulnerabilities of your plants, you can take preventive measures to protect them from common issues. This may include using appropriate pest control methods, practicing good sanitation practices, and choosing disease-resistant varieties.

3. **Maximizing productivity**: Knowing the specific needs of your plants can help you optimize their productivity. For example, certain plants may require specific pruning techniques or fertilization schedules to encourage healthy growth and higher yields. By following these guidelines, you can maximize the productivity of your garden and enjoy a bountiful harvest.

4. **Efficient resource allocation**: Researching the specific needs of your plants can help you allocate resources such as water, fertilizer, and space more efficiently. Some plants may have lower water requirements, while others may need more frequent feeding. By understanding these needs, you can avoid unnecessary resource wastage and ensure that each plant receives what it needs to thrive.

5. **Troubleshooting**: Even with proper care, plants may sometimes develop issues or exhibit signs of distress. By researching their specific needs, you will be better equipped to troubleshoot problems and identify potential causes. This knowledge can help you take timely corrective actions and prevent further damage to your plants.

Overall, researching the specific needs of the plants you want to grow is crucial for providing them with the best possible care, preventing problems, maximizing productivity, and ensuring the success of your garden.

Caring For A Garden

Caring for a garden involves several key tasks to ensure its health and beauty. Here are some essential tips for garden care:

1. **Watering**: Provide regular and adequate water to your plants, especially during dry spells. Water deeply to encourage deep root growth, and avoid overwatering to prevent waterlogged soil.

2. **Mulching**: Apply a layer of mulch around your plants to retain moisture, suppress weeds, and regulate soil temperature. Organic mulches like wood chips or straw are ideal.

3. **Weeding**: Regularly remove weeds from your garden to prevent them from competing with your plants for nutrients and water. Use a hoe or hand-pull them when they are small.

4. **Pruning**: Prune your plants to promote healthy growth and maintain their shape. Remove dead or diseased branches, and trim back overgrown foliage. Different plants have different pruning requirements, so research specific guidelines for each plant in your garden.

5. **Fertilizing**: Feed your plants with a balanced fertilizer to provide essential nutrients. Follow the instructions on the fertilizer packaging and apply it at the appropriate times during the growing season.

6. **Pest Control**: Monitor your garden for pests and take appropriate measures to control them. This can include using organic pest control methods, such as handpicking pests, introducing beneficial insects, or using natural repellents.

7. **Sunlight and Shade**: Place your plants in areas that receive the appropriate amount of sunlight for their specific needs. Some plants prefer full sun, while others thrive in partial shade. Consider the sunlight requirements of your plants when planning your garden layout.

8. **Regular Maintenance**: Regularly inspect your plants for signs of disease or nutrient deficiencies. Remove any dead or yellowing leaves, and address any issues promptly to prevent further damage.

Remember, garden care is an ongoing process that requires attention and observation. By following these tips and dedicating time to your garden, you can create a thriving and beautiful outdoor space.

Root Cellar

Having a root cellar on your land can be beneficial for several reasons:

1. **Food storage**: A root cellar provides a cool, dark, and humid environment that is ideal for storing fruits, vegetables, and other perishable food items. It helps to extend the shelf life of these foods, allowing you to enjoy fresh produce throughout the year, even during the winter months when fresh produce may be scarce or expensive.

2. **Cost savings**: By storing your own produce in a root cellar, you can reduce your reliance on grocery stores and save money on buying preserved or imported fruits and vegetables. It can also help you take advantage of seasonal abundance by allowing you to preserve excess harvest for later use.

3. **Self-sufficiency**: Having a root cellar enables you to become more self-sufficient by preserving your own food. It gives you greater control over the quality and source of your food, especially if you grow your own fruits and vegetables. This can be particularly valuable during times of uncertainty or emergencies.

4. **Sustainability**: Storing food in a root cellar reduces the need for refrigeration or freezing, which can save energy and reduce your carbon footprint. It promotes a more sustainable way of living by utilizing natural methods of food preservation.

5. **Preservation of heirloom varieties**: If you grow heirloom fruits and vegetables, a root cellar can help preserve their unique flavors and characteristics. These varieties are often not readily available in commercial markets and can be preserved and enjoyed for generations to come.

Overall, a root cellar can provide you with a reliable and efficient method of food storage, allowing you to enjoy fresh, nutritious produce year-round while promoting self-sufficiency and sustainability.

Bunkers

Having a bunker on your land can serve a variety of purposes, including:

1. **Emergency preparedness:** A bunker can provide a safe and secure location during emergencies such as natural disasters, civil unrest, or pandemics. It can offer protection from extreme weather conditions, nuclear fallout, or other potential threats.

2. **Personal safety**: A bunker can serve as a personal safety zone, offering protection from intruders, home invasions, or other security risks. It can provide a sense of peace of mind knowing that you have a secure space to retreat to if needed.

3. **Storage and organization**: Bunkers can be used to store and organize essential supplies, such as food, water, medical kits, and other emergency supplies. This ensures that you have access to necessary items during critical situations.

4. **Off-grid living**: Some individuals choose to have a bunker as part of their off-grid living strategy. It can provide a self-sufficient living space with alternative

energy sources, water collection systems, and sustainable food storage options.

5. **Recreational use**: Bunkers can also be used for recreational purposes, such as a private shooting range, wine cellar, or a unique entertainment space. It can offer a secluded and soundproof area for various activities.

6. **Investment value**: Bunkers can be seen as an investment, as they may increase the value of your property. Some people may be interested in purchasing properties with pre-built bunkers for their own emergency preparedness or personal use.

It's important to note that the decision to have a bunker on your land is a personal one and should be based on your specific needs, concerns, and circumstances. It's advisable to consult with professionals and local authorities to ensure compliance with building codes and regulations.

Water Access

Having your own fishing pond on your land can indeed be beneficial for several reasons:

1. **Recreation and relaxation**: A fishing pond provides an opportunity to engage in a popular outdoor recreational activity. It can be a great way to relax, unwind, and enjoy nature. Fishing can also be a fun and rewarding hobby for individuals, families, and friends.

2. **Convenience and accessibility**: Having a fishing pond on your land means you have easy access to fishing without the need to travel to other locations. This convenience allows you to fish whenever you want, without worrying about crowded public fishing spots or limited fishing access.

3. **Privacy and exclusivity**: Your own fishing pond offers privacy and exclusivity. You can enjoy a peaceful fishing experience without distractions or interruptions from other anglers. It provides a more intimate and personal fishing atmosphere.

4. **Environmental control**: With your own fishing pond, you have control over the pond's ecosystem. You can

manage the fish population, control water quality, and create an ideal habitat for the fish species you prefer. This allows you to enhance the fishing experience and potentially improve the size and health of the fish.

5. **Educational opportunities**: A fishing pond can be a valuable educational resource, especially for children. It provides an opportunity to learn about aquatic ecosystems, fish species, and the importance of conservation. It can also teach patience, problem-solving, and basic fishing skills.

6. **Potential income**: If properly managed and stocked, a fishing pond can generate income through activities such as catch-and-release fishing, fishing tournaments, or even offering fishing memberships to others. This can be a way to generate additional revenue from your property.

However, it's important to note that creating and maintaining a fishing pond requires careful planning, expertise, and ongoing maintenance. Factors such as water quality, stocking fish, managing vegetation, and complying with local regulations should be considered. Consulting with fisheries experts or professionals can help ensure the long-term success and sustainability of your fishing pond.

When creating and maintaining a fishing pond, several factors should be considered to ensure its success and sustainability. These factors include:

1. **Location**: Choose a suitable location for your fishing pond that provides adequate space, access to water sources, and proper drainage. Consider factors such as sunlight exposure, proximity to trees (which can cause debris), and the overall topography of the area.

2. **Size and depth**: Determine the appropriate size and depth of your fishing pond based on your goals and available space. Larger ponds tend to be more stable and can support a greater variety of fish species. The depth of the pond should be sufficient to support the desired fish species and their habitat requirements.

3. **Water source and quality**: Ensure a reliable water source for your pond, such as a natural spring, well water, or a nearby stream. Regularly test the water quality to monitor factors such as pH, dissolved oxygen levels, and nutrient levels. Proper water quality management is crucial for the health and productivity of the fish population.

4. **Pond construction**: Constructing the pond involves excavation, shaping the pond's contours, and building appropriate structures like dams and spillways. It's important to consider factors such as soil type, compaction, and erosion control to ensure the pond's integrity and longevity.

5. **Fish stocking**: Determine the fish species you want to stock in your pond based on your goals and the pond's characteristics. Consider factors such as water temperature, oxygen levels, and the availability of suitable food sources for the fish species. Consult with fisheries experts to determine the appropriate stocking rates and species combinations for your pond.

6. **Vegetation and habitat management**: Manage vegetation in and around the pond to prevent overgrowth and maintain a healthy ecosystem. Aquatic plants provide essential habitat and food for fish, but excessive growth can lead to oxygen depletion and fish kills. Implement measures to control invasive species and maintain a balanced ecosystem.

7. **Fishing regulations and permits**: Research and comply with local fishing regulations and permit requirements. Some areas may have specific rules regarding catch limits, fishing methods, and the use of

certain baits or lures. Ensuring compliance with these regulations helps protect the fish population and supports sustainable fishing practices.

8. **Ongoing maintenance**: Regular maintenance is crucial for the long-term success of your fishing pond. This includes monitoring water quality, managing vegetation, inspecting and maintaining pond structures, and addressing any issues promptly. It's also important to periodically assess the fish population and adjust stocking rates if necessary.

By considering these factors and seeking expert advice when needed, you can create and maintain a fishing pond that provides an enjoyable and sustainable fishing experience.

Transitioning to a rural, off-grid lifestyle is a significant step that requires careful planning, determination, and a positive mindset. By following the outlined steps—from securing permits and utilities to managing the challenges of packing and relocating—you can lay a solid foundation for your new life. Embracing sustainable practices, leveraging resources from the USDA, and cultivating a mindset of abundance and purpose are key to making your dream a reality. While the journey may come with its share of challenges, the rewards of independence, sustainability, and a deeper connection to nature are well worth the effort. Remember, with the right preparation and a belief in your vision, your dreams are within reach. Step confidently into your purpose, and let this guide be your companion on the path to realizing your rural living aspirations.

About The Author

Imecca Wright is a dedicated mother, grandmother, and family-oriented individual who understands the importance of strong family bonds. With a passion for preserving family traditions and creating cherished memories, Imecca is a natural matriarch and leader within her community.

In addition to her role as a family advocate, Imecca is a Certified Life Coach and Certified Herbalist with over 30 years of experience working in the mental health field. Her personal experience with trauma and loss has inspired her to make a positive impact on the lives of others.

After experiencing the devastating loss of her son and his brother, Imecca found the strength to rebuild her life and help others in the process. She founded the Wylderness program, a youth mentoring and counseling program that aims to reach children at a young age and provide them with the tools they need to navigate life's challenges.

Imecca's dedication to her family and her community is unwavering, and her passion for helping others is truly inspiring. Her story serves as a testament to the power of resilience, faith, and determination in overcoming adversity and making a positive impact in the world.

Land Your Dreams Volume Two

I. Wright

Notes

Land Your Dreams Volume Two

Notes

I. Wright

Notes

———————

Land Your Dreams Volume Two

Notes

I. Wright

www.ingramcontent.com/pod-product-compliance
Lightning Source LLC
Chambersburg PA
CBHW070338230426
43663CB00011B/2366